I0796587

EX LIBRIS

For My Unicorns

—

A group of unicorns is called a blessing,
and that checks out.

A Sleuth of Bears

and Other Amusing, Beguiling, and Peculiar Collective Nouns

Colter Jackson

PA PRESS
PRINCETON ARCHITECTURAL PRESS · NEW YORK

The animals are busy, as you will see.
The animals are busy, on land and at sea.

A bed of sloths keep cozy and warm...

while a band of gorillas
play in a storm.

THE
GORILLAS

A bloat of hippos gather around...

and a shiver of sharks don't make a sound.

A sleuth of bears try to figure it out...

while a prickle of hedgehogs roll about.

A tower of giraffes tumble and sway...

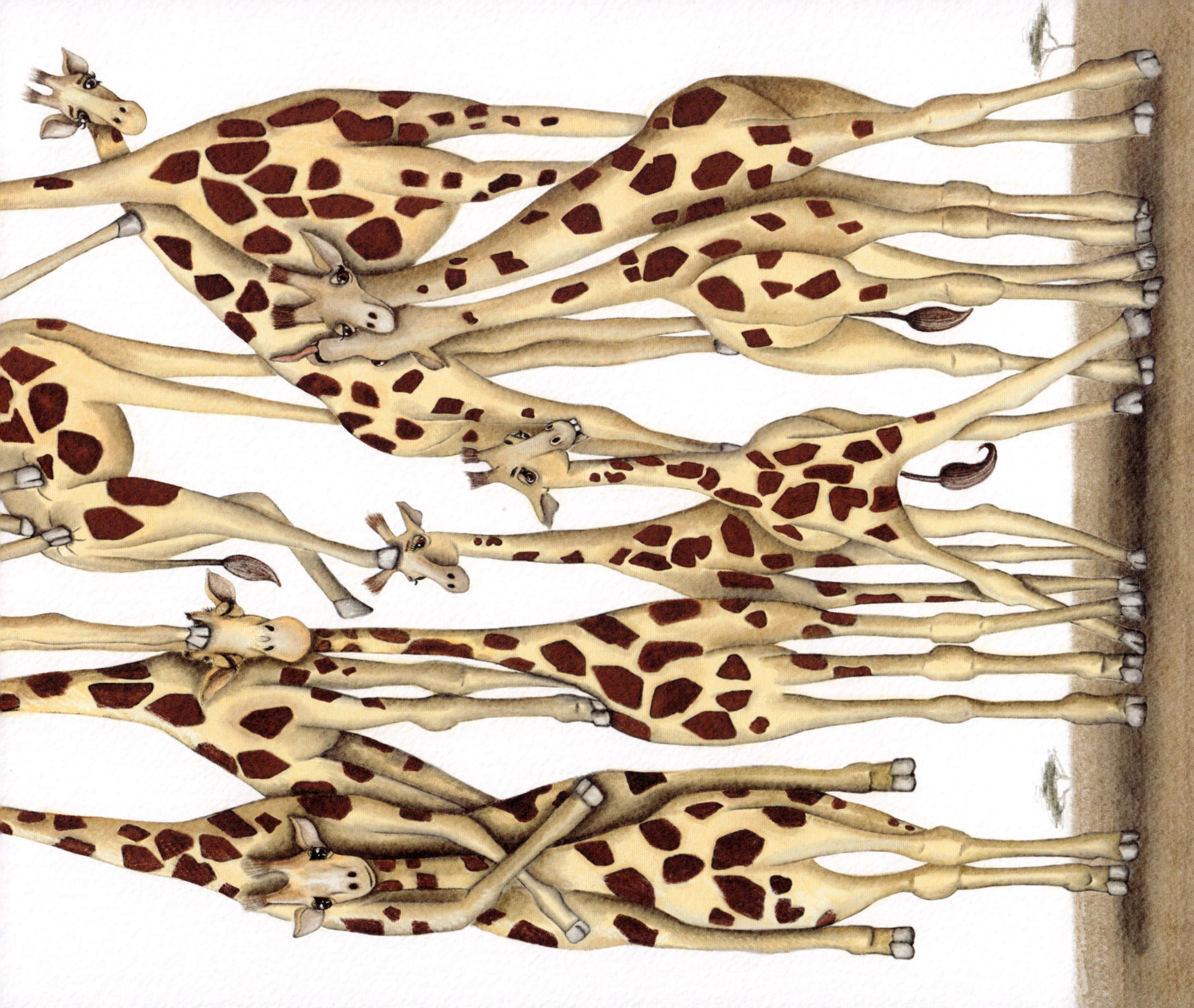

while a scurry of squirrels hide nuts away.

A memory of elephants go marching by...

while a parliament of owls stand judging nearby.

The animals are busy, all night and all day.
The animals are busy—get out of their way.

A cloud of
bats darken
the sky...

while a flamboyance of flamingos
go skating by.

LOVE

A crash of rhinos cause a delay...

while a romp of otters
all play croquet.

An ambush of tigers fall asleep
while they wait...

and a creep of tortoises try not to be late.

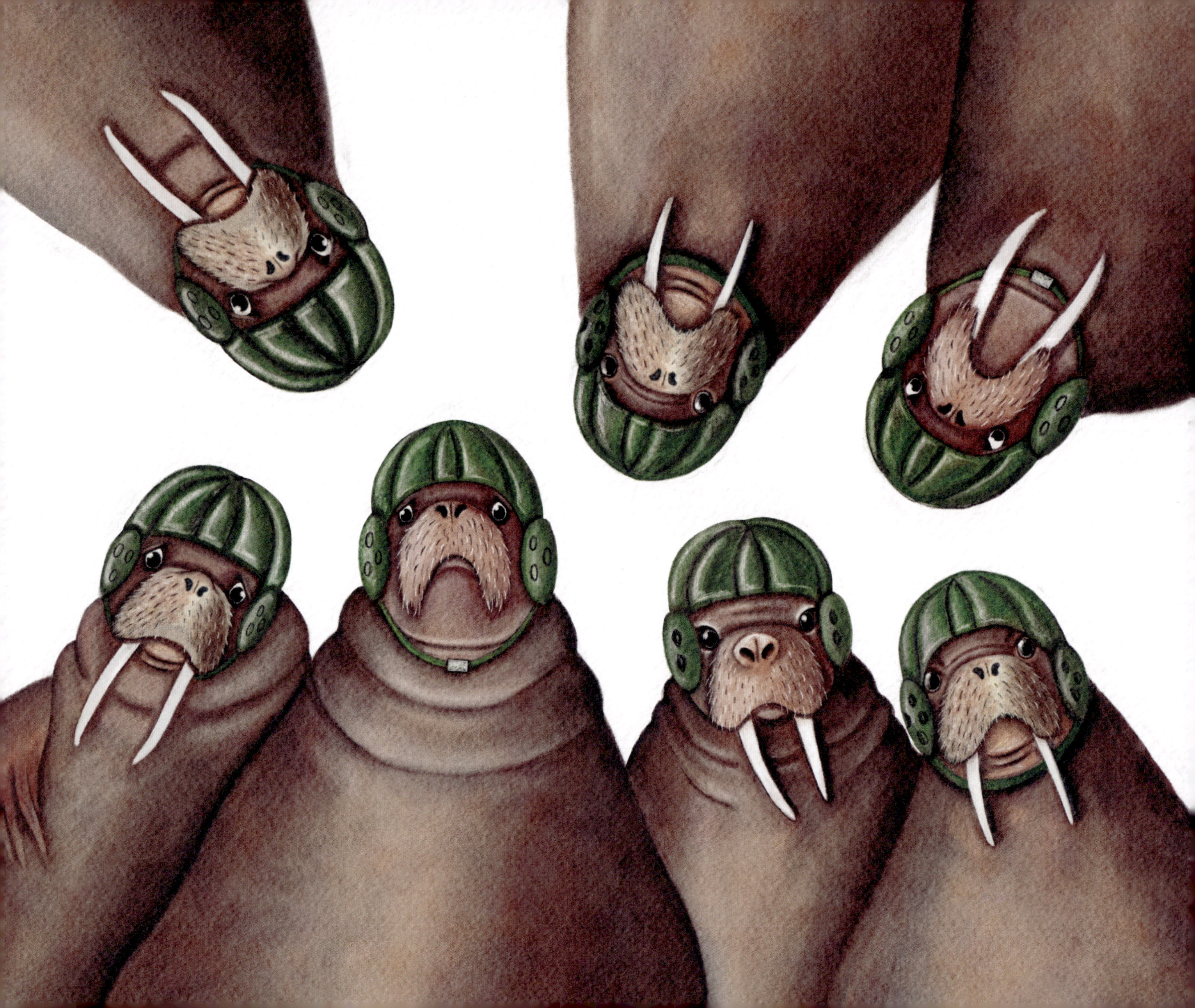

A huddle of walruses go over the play...

while a clutch of chickens lay eggs in the hay.

A congregation of gators all sing a hymn...

while a dazzle of zebras refuse to blend in.

The animals are busy—it has always been so.
The animals are busy, above and below.

A murder of crows
lurk on a limb...

while a grumble
of pugs head
to the gym.

TEA
TEA
TEA

A colony of chinchillas throw out all the tea...

and all the while, a pod of whales are out to sea.

The animals are busy—we know this is true.
The animals are busy. How about you?

Published by
Princeton Architectural Press
A division of Chronicle Books LLC
70 West 36th Street, New York, NY 10018
papress.com

Printed and bound in China
28 27 26 25 4 3 2

ISBN: 978-1-7972-3303-1

Editor: Jennifer N. Thompson
Designer: PA Press

Library of Congress Control Number: 2024036052